Cruisers

Cruis

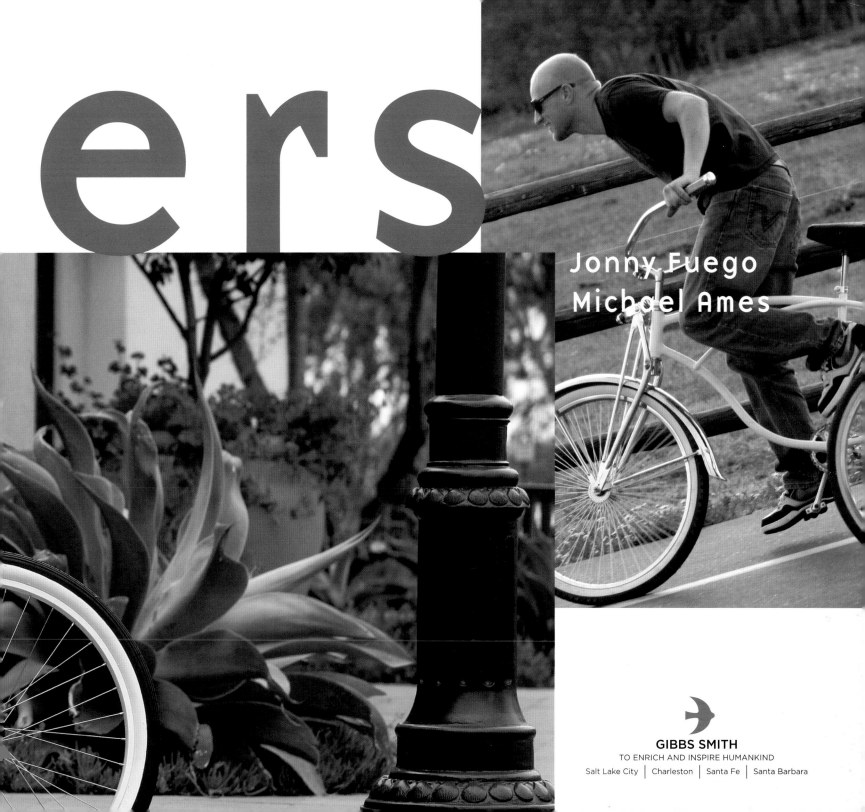

ers

Jonny Fuego
Michael Ames

GIBBS SMITH
TO ENRICH AND INSPIRE HUMANKIND
Salt Lake City | Charleston | Santa Fe | Santa Barbara

First Edition
13 12 11 10 09   5 4 3 2 1

Text © 2009 Jonny Fuego
and Michael Ames
Text on page 138 excerpted
from Katy Dang, "Ready, Set,
Race," originally published
in *Boise Weekly*, October 26,
2005. Reprinted by permission
of Sally Freeman, *Boise Weekly*.

Published by
Gibbs Smith
P.O. Box 667
Layton, Utah 84041

Orders: 1.800.835.4993
www.gibbs-smith.com

Designed by Kurt Wahlner
Printed and bound in China

Gibbs Smith books are printed
on either recycled, 100% post-
consumer waste or FSC-certified
papers.

Library of Congress Cataloging-
in-Publication Data

Fuego, Jonny.
  Cruisers / Jonny Fuego and
Michael Ames. — 1st ed.
    p. cm.
  ISBN 978-1-4236-0267-5
  1. Road bicycles—United
States. 2. Bicycles—Collectors
and collecting.
3. Machinists—United States—
Biography. I. Ames, Michael.
II. Title.
  TL437.5.R63F84 2009
  629.227'2—dc22
                    2008037618

Contents

Preface

I blame it on a *Sting-Ray.*

When I was eight years old, I was way too small for the 26-inch cruisers splashed across the pages of the old 1950s magazines that my parents kept. I let them know how much I dug those bikes, so for my eighth birthday they bought me an old Schwinn Sting-Ray. It was red with a sparkled banana seat and riser handlebars. But I grimaced at the sight of training wheels. My father had a solution. I watched as he turned the bike upside down, got out his wrench, and began disassembling the rear wheel to take off the humiliating trainers.

That was the moment I realized that with the wrench and common tools in my dad's rusty red toolbox, I too could remove parts and replace them with whatever I could find. That's where it all began.

Kids who lived on my street were into BMX racing, and some of them were even into motorcycles. Naturally, I wanted a bike that looked like theirs. The bigger kids had a lot of used parts lying around their garages. They kicked down a few to me and soon I had my own small collection. Jeff gave me a set of yellow mag wheels and beat-up motocross handlebars. From Darren, I got a black hard plastic seat with a chunk missing from the back, the result of crashing hard to the pavement after bunny hopping a trash can. From other neighborhood kids, I scored a gold-fluted seat post, a set of gold pedals, and some worn knobby tires. With tools in hand and my bike upside down, I found that the original parts came off pretty quick, and the new parts went on just as fast.

When my parents saw what I was doing to my bike, they flipped. To them, I had destroyed my birthday present. My father ultimately gave in and said, "Well, since you've come this far with all these new parts, we should probably just repaint the whole thing." We picked up a few rattle cans of spray paint and started taking my bike apart again. Soon enough, with the parts dangling like wind chimes from strings off the old tree in our yard, I began painting them all a deep matte black, except for the gold seat post, pedals, and the mags.

That was my first custom bike.

December 24, 1978, is a day I will never forget. My family was on Christmas break. After I rode home from a friend's house, I locked my bike to the trusty backyard tree. We had dinner and watched a show on TV called *Kiss Meets the Phantom of the Park*. As I sat down to watch, I looked out our back window. My bike was gone. "Damn it," my father said under his breath. He had bought me new tires for Christmas.

The memory of that bike is always with me. Not because it was my first bike—it wasn't. It stays with me because it was the first bike I built.

Today, when I see a bike on the street, I can't help but see how this machine could be improved and personalized for its owner. The purpose in having a custom bike is to make it yours. The aim of this book is to inspire. We should not accept what is suggested for our bikes, but instead go out there and make them our own. I want everyone who reads this book and who loves bikes to feel what I felt about my first custom. I want each of you to be on a bike with style, a bike that says who you are. Don't be scared—anyone can do it.

—Jonny Fuego

Introduction

A bicycle is a simple machine. Two wheels, handlebars, and a frame are about the sum of its necessary parts. Although it has undergone countless innovations, enabling smoother motion over varied terrain, its simple purpose remains unchanged. The bicycle's function is definite.

Form, however, is another story.

Sometimes getting from point A to point B, day after day, gets a bit tiresome. The classic American cruiser was devised as an antidote to such monotony. Introduce bold style to a machine and it becomes an experience. Enshrine that experience in the nostalgia of halcyon days in 1950s California, and you have an icon.

The modern cruiser revival began with the new millennium. After generations on the sidelines, the cruiser bike was ready for a comeback. It all began again in Southern California, but stylish bikes soon spread to beach and mountain towns across the land. By the time they reached the suburbs, the resurrection was complete.

Yes, the modern cruiser movement is about style and comfort and memory. But ultimately the cruiser offers something more. Low-slung and dripping with stylish sensibility, it makes a grand offer. Modern lives are complex and competitive, harried and programmed. The cruiser strikes a deal: get out of the car, hop on the saddle, and feel the easy wind in your hair. Rediscover the smooth rolling freedom of your lost childhood.

The Classic Cruiser Clique

Classic cruiser collectors cherish their bicycles. They pursue the hobby with a patriotic zeal. And as with flags flown in early wars, the origins and details of these bikes are often in dispute.

It's not just the roots of vintage cruisers that are parsed; it's any and every facet of these classic machines. So close are they to the soul of an original American culture that, like with jazz or baseball, the details that would strike the layperson as hopelessly minute become the fulcrum upon which serious collectors rest their knowledge.

In the small world of classic bicycle enthusiasts, collectors become connoisseurs and connoisseurs become experts. For the last fifty years or so, these experts worked mostly alone,

1939 Schwinn DX "Lasalle"

The head badge on a bike showed what kind of bicycle it was, but not necessarily who made it.

Huffman Firestone Twin Flex 1938

This classic has both front and rear suspension.

Store owners held award ceremonies in front of their doors to draw in crowds.

Schwinn DX from an old magazine advertisement.

tinkering in backyard sheds and garages, dusty workspaces littered with decades of rusty parts, yellowed catalogues, and boyish obsessions with perfect toys. These men (one finds few women in classic bicycle circles) would gather at seasonal swap meets and search for just the right spring-loaded fork or reflector casing. Information—about where to find a particular prewar part, for instance—was secretive and exchanged only in person or through small newsletters. The hobby was a decentralized cult; a diaspora of obsessives spread over the continent and around the globe.

For a half-century, this academic exchange of knowledge held steady. The Internet, as is its tendency, changed all that.

This ad not only shows new bikes but also the accessories you could slap on.

Offering more accessories than the "other guy" gave bike shop owners the edge over the competition. Offering a layaway plan sweetened the deal.

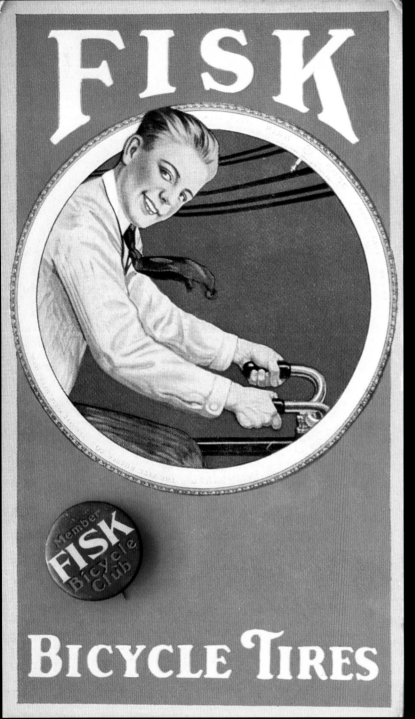

Advertising at its height in car and
sports magazines.

A front mount light and horn combo made to resemble the big touring motorcycle of the day.

1916 Hendee Indian

A sweet fender skirt.

A kickstand connection with rolled spring sits on the rear wheel.

Today, loyalists gather on Web sites like bicyclechronicles.com, nostalgic.net, and classicrendezvous.com. Here they trade, inquire, boast, and fully immerse themselves in classic two-wheelers.

"The Internet brought everyone onto a level playing field," says Tim Brandt, founder of Bicycle Chronicles. Thanks to the open sharing of information, particularly in regard to the cost of old bikes and their parts, even novice collectors today stand a better chance of scoring a fair deal, he says.

One man's junk is another man's treasure, and when it comes to classic bikes, the difference can often be several thousands of dollars. In the old days, if a couple of hayseeds were cleaning out a barn, there was no way for them to know whether their dinged-up Schwinn might be some California collector's dream.

⬆ ⬆

The tank resembles that of an
Indian motorcycle. Very cool.

⬆

Simple stem design.

➷

The whole beautiful package.

Defining a Classic

Leon Dixon, founder of the National Bike Historical Archives of America (NBHAA), is the authority on classic bicycles. After forty years of research and methodical collecting, Dixon has amassed an unparalleled archive. He claims three hundred original bicycle films and sixty thousand original catalogues and advertisements for a century of bicycles beginning in 1860. Widely regarded as the first expert to define the classic era, Dixon wrote defining articles in *Classic Bicycle & Whizzer News* in 1978 and again in the November 1979 issue of *Bicycle Dealer Showcase*. In the latter article, Dixon defined a "classic" as any bicycle built between 1920 and 1965, which includes a variety of types, not all of them cruisers.

While never authoritatively defined by anyone other than Dixon, it may safely be said that a classic cruiser is any bike from this era built with some common cruiser elements such as balloon tires, wide handlebars, a comfortable and wide saddle, and some amount of stylization.

A lot of different brands of bikes were made by the same bike manufacturer. Oil companies, tire companies, and anyone with an advertising budget could ask a major bicycle company to build a bike and create a head badge for their company.

"Some collectors say the Internet hurt the hobby," Brandt says. But he figures that the cranks issuing those complaints are just incurable nostalgic-types pining for the glory days when stumbling across a grossly underpriced classic was common. While the information superhighway has made it easier for newcomers, there are still plenty of obscure bikes out there and even more out-of-the-way places to find them. Internet or no Internet, little can discourage the serious collector.

But what qualifies a serious collector?

"If you are willing to get on an airplane and fly across the country to go to a swap meet, you are a serious collector," says Dave Stromberger, who has flown from his northwest home near Spokane, Washington, to several meets in Michigan and Ohio. Stromberger is the man behind nostalgic.net, one of the largest online hubs for classic collectors of all brands and pinstripes. Each year, he heads to the Midwest for big spring meets hosted

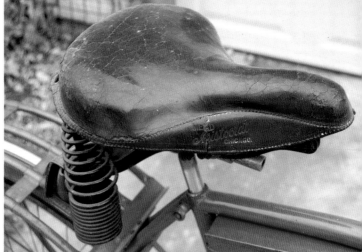

Head badges, advertisements, and accessories: bike companies knew what it took to sell a bike and retain customer loyalty.

Overland badged, Rollfast-made bike.

Must-Have Classic Schwinns

Tim Brandt, prewar Schwinn expert and the founder and editor of bicyclechronicles.com, shares his top four prewar Schwinns:

1938–1940 Hanging Cantilever Autocycle

The '38 model is the most desirable, as it was the first to feature a double-duty fork and aluminum fenders. "Most argue that the Aerocycle is more rare, but I beg to differ, and I am a prewar Schwinn expert."

1934–1936 Aerocycle

It was voted the world's greatest bicycle in 1935, but Brandt is skeptical of the overall quality. "I believe they were voted as the best for their appearance only, because the frames were notorious for breaking."

1936–1938 Jewel Tank Autocycle

The '38 model has the most features.

1935 Double Diamond Cycleplane/Motorbike

↘

1938 Schwinn Autocycle

There are less than ten known originals of this jewel-tank model.

↑

1938 Schwinn Cantilevered Autocycle

A super-rare find—the best-known original of this cantilevered autocycle in existence.

⟳ ⟱

*Huffman Firestone
Twin Flex 1938*

A detail shot of the complex rear
suspension. It pivoted at the rear
dropouts.

↑ →

1941
Straight-
Bar Schwinn
Excelsior

The owner bought this bike to be ridden and didn't want to splurge on paint.

by Memory Lane Classics. There, he meets up with specialized brand collectors. "Every brand has its coolest collector model," he says. Stromberger is the outsider foil to Brandt's Schwinn brand loyalty.

By day, Brandt is a detail-oriented accountant, but he has a long history in the bike biz, first as a mechanic in a shop in his home state of Wisconsin. After graduating college, he "just wanted a cool bike to ride around on." Around that time, he found his first prewar Schwinn. After he moved to Southern California, the hobby became a second business. He went to local swap meets and met collectors who would fly to scout out Midwest bike shops in the hope of finding old parts that could be bought on the cheap and resold at a premium. Some of those shops have since caught on to the California collectors' schemes.

"Once word got around that this stuff was valuable, bike shops became much more apprehensive," Brandt says. When he goes on classic bike quests nowadays, he is met with more wariness. "When you walk in and ask if there's anything old for sale, a wall goes up."

Through bicyclechronicles.com, Brandt sells old parts on eBay. "We start things at $9.99 and the best man takes it." The way he sees it, fair pricing is a way to give back, to circulate parts among a community that truly enjoys collecting. That community, it turns out, is quite small—a "close-knit group," Brandt calls it—and it's not unusual for a single part to go back and forth between the same few people. "My girlfriend said it's the most incestuous hobby she's ever seen."

Beyond the Web site sales, the bulk of Brandt's passion is in restorations. This is arduous, time-consuming work; a full

→

1941 Columbia Westfield

Simple earthy colors.

1948 Schwinn Hornet

Creative stampings and paint scheme marked a good chain guard.

Non-Schwinn Classics: The Best of the Rest

While prewar Schwinns continue to dominate the classic market, there is a sizable inventory of non-Schwinn classics that not only appeal to their own legions of admirers but also fetch equal money, and sometimes more, at swap meets.

1935 Elgin Bluebird

Elgin was the brand manufactured by Sears, Roebuck and Co. during the mid-to-late 1930s. The Bluebird was heavily streamlined and styled for its time and was not lacking in nifty bells and whistles. It came equipped with a lit speedometer and headlight, a battery compartment, and one of the industry's earliest kickstands. But with the Depression on, Sears did not do very well with their hyperstylized Bluebird series. Sales were low and production decreased, making the Bluebird one of the most elite classics available today. Even an unrestored model can pull upward of $10,000. In top restored condition, Elgin Bluebirds are among the most valuable classic bicycles in the world.

1938–1940 Shelby Speedline Airflow

Competitive with contemporary Schwinn and Elgin, this company out of Shelby, Ohio, launched the Airflow series, chock-full of design firsts. The 1939 model had an electric horn (operated with a button on the tank) and a rear taillight.

Roadmaster

Produced by the Cleveland Welding Company, the 1937 Supreme is the most sought-after model.

Hiawatha

Also a Cleveland Welding Company bike, Hiawathas were sold by Gambles Hardware stores. Some were also made by the Shelby Manufacturing company. The late-1930s Arrow-model Hiawatha is prized for its extreme styling and resemblance to the Shelby Airflow.

paint job takes him roughly forty hours. Lucrative it is not.

"Sometimes, I do question the amount of time I put into this. It can spill over and affect your personal life. My friends and girlfriend make comments." He tries to prioritize his time, and, despite the occasional conflict with the nonmachines in his life, Brandt's simple passion has not abated.

"I still have a love for the bikes. They still get me excited and get the endorphins popping."

And when he finds a true gem—an unclaimed classic Schwinn needing the love, knowledge, and care of a professional restorer—the sales figures will transcend the realm of hobby. A properly restored rare bike, a prewar Cantilevered Autocycle, for instance, can fetch over $10,000.

1922 Mead Ranger

If your bike breaks down, no worries! You have all your tools stored in the tank of this classic.

⟲ ⬆ ⟳

1950
Schwinn
Panther

The head badge
proudly shows who
made this one.

↟

1938 Schwinn C-Model "Spitfire"

➡

1948 Schwinn DX

Ready to ride.

↑ ↰

1937 Monark
Silverking
Hextube

↱

1958
Schwinn
Tornado

Fin accents give
this tank a little bit
"more."

Neighborhood kids will flock to see your new speedometer, lights, and other new accessories.

The Modern Cruiser Revival

The Cruiser's Dark Ages

In the annals of the cruiser bike, the late twentieth century was a dark chapter. By the 1970s, biking in America had become a niche activity, marketed mostly to fitness types on road bikes or to kids on Schwinn Sting-Rays. BMX bikes, like those in the classic eighties flick *Rad*, were for adrenaline-starved suburban teens. In the seventies and eighties, any notion of a cruiser was limited to classic bike collectors (many of whom didn't even ride their prize possessions) and retro-minded beach bums in Southern California. In our expanding country, cool and casual cycling was nowhere to be seen. The two-wheeled realm was ruled by athletes, eccentrics, and people who couldn't afford cars.

One of those eccentrics was Gary Fisher. Once disqualified from a track bike race for his long hair, Fisher had grown bored with road cycling's uptight culture. In need of a wild and wooly diversion, he began tinkering with a 1930s Schwinn Excelsior X. He improvised on mechanics with add-on devices like expander brakes and racing gears. Fisher took his first off-road prototype on a few harrowing downhill rides on hiking trails in the steep mountains

of Marin County, just north of San Francisco. When his friends caught on, they wanted in. In this casual way, Gary Fisher invented the modern sport of mountain biking.

By the early 1990s, Fisher's new sport was all the rage, but only for a select market. Mountain biking was and would remain an X-Games, Mountain Dew sort of pursuit—not exactly a fun and relaxing ride for your average modern American desk-jockey. Casual bike riding remained a foreign concept. Stepping into any bike shop or flipping through the pages of any two-wheeler magazine hammered the point home: cycling still belonged to spandex-clad adrenaline-freak weirdos.

Whither Rolleth the Cruiser?

Cruisers had never gone away entirely. Beyond the classic collectors and their manic perfectionism, the cruiser idea remained a cultural undercurrent. The Schwinn company knew this, but failed to capitalize.

The story of the Schwinn Bicycle Company's downfall, after decades of dominance, was material enough for a serious case study: *No Hands: The Rise and Fall of the Schwinn Bicycle Company, An American Institution*. This exhaustive book, by veteran reporters at *Crain's Chicago Business*, charts the glory-day highs and the embarrassing lows of the fallen industry colossus.

The Schwinn story is a sprawling one, but the cruiser history alone reveals both the company's poor management and how cycling in America evolved while Schwinn slept at the handlebars.

In 1980, Schwinn introduced its first bike bearing the "cruiser" badge, but it was a far cry from its classic postwar ballooners. This eighties cruiser was utilitarian, with little attention paid to the styling details that had defined Schwinn classics. It was a cheap failure of imagination.

According to Jay Pridmore, who has written several accounts of the Schwinn century, these bikes were an "unimpressive" and motley assortment of parts patched together from production facilities in Chicago, Taiwan, and Hungary. Inexpensive front forks were taken from mountain bikes, saving Schwinn money but costing riders the old easy feeling of solid-steel frames.

By the early 1990s, Schwinn's bankruptcy was in the tea leaves. To their credit, company designers powered through the insolvency swamp and came out on the other side, determined to make a new beginning.

↻ ➔

New Schwinn Retro

A new Schwinn off the floor in today's shop still has close ties to the past.

The Classic Resurrection

Amid Schwinn's sad downfall and the wholesale chaos of moving a hundred-year-old company from Chicago to Colorado, Schwinn managed to pull off a bit of magic. First came the mad search for lost grace: engineers rummaged through basement files and literally blew the dust off fifty-year-old papers. Those original postwar blueprints might as well have been mock-ups to the caveman's first wheel. Soaking in the light of day for the first time in decades, the yellowed pages breathed creative life into the company at a time when the future was never less certain.

Buoyed by their own legacy, Schwinn released its first full line of reimagined cruisers in 1994, and in doing so, officially sparked the modern cruiser revival. The bikes were instant hits. They were recognizable icons of the American Century, as indelible as Marilyn Monroe or the first Electrolux. But unlike the fame-doomed Norma Jean or the blissfully vacuuming homemaker, these were built to last.

The new Schwinns capitalized on the growing purchasing power of paunchy baby boomers and their limitless nostalgia. For the generation that grew up pedaling Jaguars, Panthers, and Starlets around bucolic tree-lined neighborhoods, the rebuilt cruiser offered a precious opportunity. By simply buying a bike, a generation could recapture the innocence of a youth lost to everything but the soundless grain and sun-bleached jumps of 8-millimeter reel-to-reel film.

In 1995, coasting on encouraging sales and determined to celebrate their centennial in style, Schwinn introduced a limited commemorative edition of the 1955 Black Phantom. That original was nearly perfect. Built at the peak of an age when form edged out function in American automobiles and bikes, the Phantom was everything a classic cruiser needed to be: comfortable, easy to ride, and gorgeous to behold. The limited edition—only five thousand were ordered—was rebuilt to exact replica detail, including integrated fender headlights, a coil-spring leather saddle, and Schwinn Typhoon balloon tires. The company confidently slapped on a $3,000 sticker price, sat back, and watched them roll.

Spreading the Cruiser Love

By returning to something with wider appeal, Schwinn not only sold thousands of Phantoms but also regained the trendsetting throne.

But retro cruiser sales were the silver lining on Schwinn's gathering gray clouds, and by the time the Black Phantom was released, the company's bankruptcy had effectively decapitated the entire bicycle industry.

Several smaller companies were quick to pick up the scent of the dying giant and flood the market with classic remakes. In the spring of 1996, *Popular Mechanics* ran a story called "Return of the Cruiser" that enumerated the speed and strength of the revival. Companies such as Columbia Manufacturing, Ross Bicycles, and Raleigh were building cruisers with a look not seen since the Eisenhower administration.

Columbia rolled out one of their most popular bikes from the fifties, the RX5, to compete with the Phantom as it had forty years prior. Western Flyer bikes joined the fracas with their Classic Springer and Classic Cruiser. Reliable Bicycle Replicas, Specialized, and Raleigh soon followed.

Some modern cruisers were simple: Raleigh's single-speed Retroglide sold at just $189 and came with few extras. At Specialized, the new cruisers were tech-minded: the Globe 7 was an easy rider equipped with modern amenities like a halogen light and an internal seven-speed Shimano gear/brake system. The mid-nineties were a tabula rasa for designers scrawling a new bike language. Cruising was coming back, and each manufacturer jockeyed to position itself as the dominant crowd-pleaser.

One such company was Southern California's Electra, a cruiser-only outfit launched in 1993 by Benno Baenziger and Jeano Erforth, two German expatriates with a unique vision of the American cycling psyche. Schwinn was on the way out, and Electra was in front of the rush to fill the power vacuum. With considerable industry inertia at its back, the young company grew into the modern era's first cruiser dynasty.

➔

Schwinn Aerosport

◉ ↑

Raleigh Classic Retro

A beautiful replica of a 1969 Tourist by Raleigh. Note the bar-rod linkage brake system, head badges, and leather saddle, all recalling the classic era.

GT Dyno
and the Custom

No comprehensive cruiser discussion is complete without mentioning the GT Dyno Kustom Kruiser.

In the mid-nineties, Schwinn, Electra, and others built nostalgic bikes with the distinct frame lines that evoked memories of classic beach bikes. But GT was the first to release a "non-beach" cruiser. The Dyno Kustom Krusier, a hot rod, broke the beach barrier in 1996.

Aaron Bethlenfalvy is a former director of design at GT and recalls turning away from Southern California and to different segments of pop culture for the Kustom Kruiser. "We took a lot of elements of the custom car culture and infused it into cruiser designs," Bethlenfalvy says. "Flat-land Middle America is going to car shows and people are customizing their bikes to make them look like hot rods or motorcycles," he says. "We took that concept and put in a box." Within a few years, GT's hot-rod bikes had been adopted by several competitors.

When he reflects on his career, Bethlenfalvy sees the Kustom Kruiser as a bright spot. As a designer, his goal was to change product perception. "It's one of the hardest things to do," he says, "and the Dyno Kustom Kruisers did that. It changed the perception on what had only been beach bikes. I'm proud of that."

For his deserved pride in his work, Bethlenfalvy is not bitter that his designs were the basis for countless rip-offs. He has a serene outlook on how design improves people's lives. "The real winner is the consumer."

The Casual Electra Age

Baenziger and Erforth moved to the United States determined to give back to America a vital piece of culture they felt it had lost. Baenziger had a background in graphic design, and, after working at companies like Adidas and K2, set out to carve his own niche.

"The cruiser was a language that I knew people would understand," he says. Having lived in Berlin, a city where bicycles are practical everyday tools—for commuting, errand-running, or just simple enjoyment—Baenziger was troubled by the lack of casual cycling in America. It had not always been this way. Sifting through fifty-year-old magazines, he realized that, at one point, Americans saw themselves on bikes. But the only cruisers Baenziger saw was in old reruns, "like *Lassie.*" Somewhere along the line, America had lost sight of its own icon.

But Electra's philosophy was not grounded in the visual candy of cruisers. Anyone could produce that.

"Everybody loves cruiser bikes," Baenziger says. "Like they love puppies or hamburgers or apple pie." He was attracted not so much to the easy love, but to the more fundamental issue of the lifestyle. His vision was a stark contrast to the type of sport cycling dominating the market in the early nineties. He was not interested in how many miles a road cyclist covered in a weekend or how many vertical feet the mountain biker had climbed.

"What about regular people riding regular bikes? What happened to that?" he wondered.

Part of the problem was marketing. It was easy to turn people on to mountain biking with splashy advertisements showing awesome feats of daring: mad men in helmets jumping off cliffs and riding roughshod down shale-covered slopes. Mountain bikers had

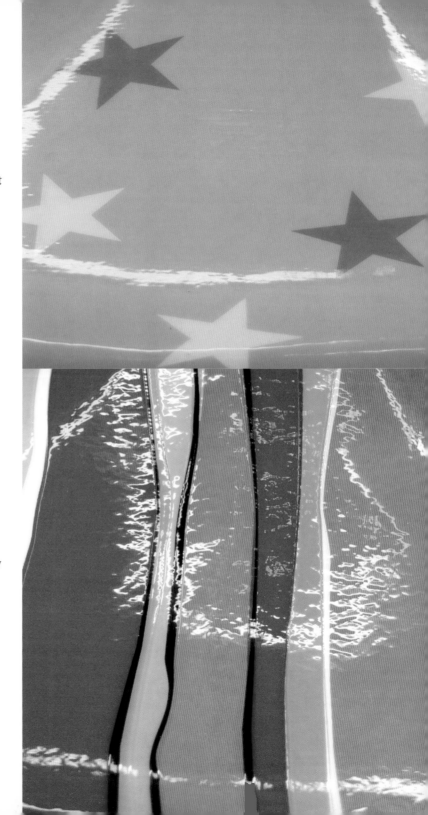

the X-Games and road bikers had Lance Armstrong. But how do you visually attract people to riding a bike to the store for a loaf of bread and a quart of milk?

Electra started by capitalizing on the basics: cruiser love. Cruiser love can be defined as an irrational exuberance that overcomes even the most ardent opponents of fun. Baenziger knows cruiser love well: "I've seen it a million times—everyone who jumps on a cruiser can't stop themselves from smiling."

Starry-eyed bike romance was a cornerstone, but when Electra needed a business model, the company turned back to Europe and Switzerland's Swatch watch. The watches were primarily functional; they told time. They did this reliably and inexpensively. But what separated Swatch from the pack was an unprecedented diversity of styles. There was a different Swatch watch for every person: the funky artist, the professional athlete, even the little girl in pink. Each unique customer could find a style ready-made to fit his or her own life.

Electra pursued this ambitious democracy of style and introduced a slew of models. These fashionable options were aggressively marketed alongside the lifestyle concept. The bikes invited customers to swagger, strut, and pedal their stuff.

"We call it the one-man parade," Baenziger says.

This combined promise—you can return to the simple pleasure of riding a bike and define your individuality along the way—catapulted the company to success. Between 2001 and 2006, Electra's revenue quadrupled.

◔ Electra combined paint and seat fabric diversity like no brand had done before.

Electra Bikes

Ladies' Classic 3 Amsterdam

A rebuild of a classic European city bike.

Betty

A basic Electra girl's cruiser.

Gypsy

A bike that is sure to appeal to the girl who wants a bike from the past, today.

Straight 8

Eight ball wins the game.

Super Deluxe ↘

Electra uses retro styling with
modern-day features.

Felt El Guapo

Felt Hurley PR ↘

Dawn in SoCal

The modern cruiser boom belongs to the twenty-first century. While Schwinn, Electra, and others were softening up the market in the nineties, cruiser fever wasn't truly epidemic until the new millennium. And it wasn't for another five years or so that people outside of Southern California started paying attention.

In the summer of 2005, the *New York Times* ran a belated story about the cruiser revival. The speed and volume with which these bikes were selling was astounding, and a large part of the story was the business angle. At Santa Monica's Bike Attack shop, the *Times* reported sales of one hundred to two hundred cruisers per month for five consecutive months. In October 2004, trade magazine *Bicycle Retailer and Industry News* picked up the story and dug up even more head-turning numbers: in the first half of 2004, cruiser shipments had increased 43 percent over 2003. In one year, sales had risen by more than half.

More media attention soon focused on fashion-minded co-branding taken up by companies like Nirve, who linked design arms with Paul Frank, Hello Kitty, and even the fashionably unfashionable John Deere. Suddenly, bejeweled young city girls were taking to the streets with cruisers trendy enough to match their accessories.

The change in habits had a lot to do with a change in industry priorities. Graphic designers and artists gained on mechanical technicians. For the first time since the postwar salad days of '57 Chevys and Schwinn Black Phantoms, style meant something in American bike engineering.

Southern California became a petri dish of cruiser innovation. Designers had sudden power. At companies like Felt, Nirve, Electra, and Phat, creative departments became driving forces. And when creative professionals switched companies (or formed their own, like Gary Silva at 3G), the industry benefited from the cross-pollination. Young designers breathed hip visual elements into companies that for decades had been focused on hi-tech hardware.

Brett King, a former BMX racer and custom helmet designer from Vancouver, Canada, was one of these creative agents. King was at GT bikes before being hired by Felt president Bill Duehring in 1999.

⟳

Felt Nectar

↻

Felt MP Rat Patrol

"Here we are, a few years later, kicking ass," King said at his Felt offices, where his design team developed a successful mix of Nirve's big-name co-branding and Electra's Swatch-inspired diversity. In 2008, Felt sold twenty-five different cruiser designs, far more choices than Electra, the company responsible for creating the market demand in the first place.

It's easy to tell when a business encourages its creative departments. When Felt's dreamers romped in the workplace, consumers were the ultimate beneficiaries. Brett King's cruiser designs can be traced back to the Kilroy, a bike modeled on early World War II–era motorcycles. Army drab paint and some dog tag stenciling was all it took to transform a regular old cruiser into the Kilroy "army bike." He then designed the MP, another army-themed ride with a more extensive motif. The bike came equipped with a canvas saddlebag and an oversized fender-mounted headlight.

At Felt, the sky was the limit. Can we build a vintage motocross cycle with a high front fender and race-number badge? Sure! How 'bout a checker cab bike with yellow-and-black detailing? Why not?

In the past, ladies' bikes had a low-slung crossbar for riding with a dress. Today shorts are in style, but the swooping crossbar remains.

Felt Hot Wheels® Sixty-8

Felt Hurley PR

Surf City USA: The Town, the Bike, the Myth

Trendy fashion sells fine, but the essence of the cruiser revival was always rooted in the power of nostalgia. This is where Huntington Beach, California, enters (or re-enters) the cruiser picture.

Felt and the City of Huntington Beach teamed up in 2005 to market the Surf City USA cruiser, a bike loosely based on a prewar Schwinn. Felt approached James Traub, president of the Huntington Beach Conference and Visitors Bureau, to see if the town would be interested in having its own custom bike. For the bureau, which had just completed a city branding project, it was instant corporate love.

"It was a cosmic marriage," Traub recalls. "We had just defined a brand that had defined the entire Southern California beach lifestyle."

One of Felt's first co-branded models, the Surf City USA cruiser marked the intersection between Southern California's mythical past and its soul-reviving present.

In 1963, Jan and Dean hit number one on the Billboard charts with "Surf City." The same year, the Beach Boys' surf homage, "Surfin' USA," topped out at number three on the Billboard Charts. Brian Wilson, lyricist to

Felt Surf City USA

Huntington Beach, California teamed up with Felt to capture the feeling of where it all began.

both songs, was creating sounds that would define a generation. The songs celebrated a wholesome, though increasingly free, youth culture. Surf City wasn't a real place, but a mythical spot on the map somewhere between Malibu and La Jolla. You could park your woody station wagon by that beach, ride the shimmering waves, and light a bonfire. On that beach, there were "two girls for every boy."

Thirty years later, with his song as scripture, Dean Torrence moved back to Huntington Beach. He reflected fondly on the place that he helped define and eventually teamed up with members of the visitors bureau to brand Surf City USA. In 2004, Huntington Beach filed with the U.S. Patent and Trademark Office. Bettering the Big Apple and the Big Easy, Huntington Beach–Surf City USA had trademarked its own nickname.

Huntington Beach is a real place, a city of bankers, bus drivers, and waitresses. But when this laid-back and temperate town moved to define its own niche mystique, city elders knew just where to turn: the cruiser bike. In Surf City USA, the ultimate California icon still holds the two-wheeled promise of eternal American youth.

Felt Heritage

Industry Killed the Custom Cruiser Star

There are downsides to flexing your creative muscles for a big bike company: for example, helplessly watching as your once-cherished personal bike design is hijacked for profit. The corporation—even a funky cruiser bike corporation—owns the creativity happening within its heartless domain. The corporate designer's one-of-a-kind bike is suddenly within reach of any yahoo with a charge card.

Brett King learned this lesson the hard way. His most-prized custom cruiser, lovingly pieced together in his own garage shop, was called the Little Bastard. King modeled it after James Dean's infamous Porsche 550 Spyder, the tool for his last rebellious act.

"The bike was just awesome," King says. His bosses at Felt agreed, and within a year, King's custom was just another barcode in the Felt catalogue.

King mourned: "I wanted that bike for me."

King's frustration over losing the Little Bastard highlights the flimsiness (and the infinite value) of individuality in cruiser culture. His angst is both universal (How do you identify yourself?) and specific to the style-driven bike company (How do you mass-market the promise of unique identity?). Everyone wants to be unique, but at the end of the day, how proud can you be of your one-man parade when countless other dudes and dudettes are marching to the same quirky beat?

↑
The Little Bastard design, after Felt got its hands on it.

➔
Waves of cruisers at every bike shop as you walk in.

The Fourth Gary

Aside from their love of bikes, three of the most influential American bike designers have one thing in common: all of them are named Gary. Gary Silva named his bike company 3G in honor of the Garys he admires: Gary Fisher (mountain bike inventor), Gary Klein (Cannondale welding innovator), and Gary Turner (GT Bicycles founder).

Silva has done well to place himself in the Gary lineage. He used to ride his cruiser around Bayonne, New Jersey, a tough dockside town where his dad owned a bike shop. From the grit of Jersey, Silva headed for warmer climes and moved to south Florida, where he opened his first shop, Gary's Megacycle. The one shop soon grew to eight large stores stretching eighty-six sandy miles from Miami clear to Juno Beach.

In 1995, Silva started "messing around" with choppers—piecing together random bike parts to create original geometric designs. In 1997, Silva closed up his Florida shops and headed to Southern California. There, in the bike design petri dish, he met the people (like Canadian graphic designer Liam Hayes) who would form his team when he helped launch Phat Cycles. Since then, Silva has devoted his life to bike building; he even lived in Taiwan for six years to further embed himself in the manufacturing process.

In 2003, Silva founded 3G. Silva's geometric frame lines have found their purest expression in 3G bikes. His lines are matched by a extreme versatility. Silva built a single frame designed to be as comfortable for the petite woman as the burly linebacker. He achieved this unusual degree of flexibility by incorporating modular handlebars with exceptionally long adjustable seat posts. In this chapter of Silva's journey, versatility has become a core belief, but frames are where he has left his mark.

As Silva says, "I take pride in my geometry."

3G bikes with Silva's signature lines.

2008 3G Stepper

This vexing problem lies at the core of the custom movement, a trend whose strengths owe much to the cruiser's successful commodification. People have been tricking out bikes for decades, but today's backyard builders are challenged to break ever more barriers. The future's cutting-edge cruisers will not be found in glossy catalogues or on bike-store racks. The new signifiers of creative struggle will not have price tags hanging off the handlebars.

Around the globe, the underground custom movement is spreading. But ascent in this world is not straightforward; this is not corporate America. As a builder gains a reputation and a client base, he must approach popularity cautiously, lest he squander his underground cache. How does the garage-bike genius balance romantic creative expression with the monthly rent? For most, the problem is solved with simple bartering: a case of beer for some new ape-hangers, a favorite bottle of hooch for a sick new sprocket. Within this simplified economic system and its quiet rejection of corporate values, the custom community is setting its own priorities.

➔

The idea, expertly marketed by bike companies, is that a cruiser bike can make you as tough as this guy.

In Hailey, Idaho, Laura Higdon found her orphaned custom at an antique sale.

Customs

For the yearning masses striving to be free of corporate definitions of cool, there is the siren call of custom bike building. Customs come in many shapes and forms. And the tribe of people that builds them are as diverse as any in America. They are doctors and lawyers, cops and plumbers, professional cyclists and artists. All of them share a similar passion: a love for bikes and a desire to create unique machines. Shelling out hard-earned cash for a mass-produced cruiser is simply not an option for these types.

The novice might start with some casual searches in the classifieds, but easily progresses to rooting around the town dump for hastily trashed frames. Mechanical and style adjustments come next—stripping rusted paint and spraying on a fresh coat, or ratcheting some big ape-hanger handlebars and a leather saddle onto an old town bike. Or hooking up a retro wire basket to an '88 Peugeot and replacing those road slicks with some fat balloon tires. Whatever makes

➔

Chupacabra Chopper

Every detail of this chopper is handmade or fabricated specifically for this bike. This is a true "one-off" design.

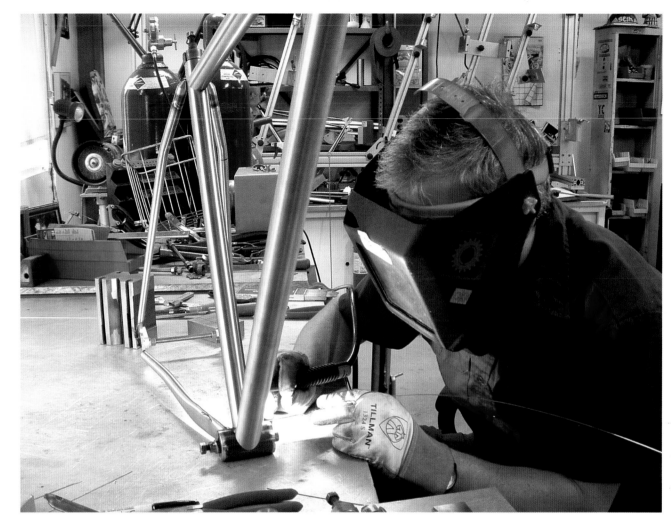

the afternoon grocery store runs or the ride home from the bar a little easier to navigate.

But for some enthusiasts, these customizations are superficial stunts. A professional builder starts from scratch; nothing is purchased, found, or remodeled. These Dr. Frankensteins of the cruiser world are creators. For them, there is no task too arcane, no skill too mechanized. If the bike can be imagined, it can be built.

◒

Tinker, tailor, solder, spy—this is where it all happens: Jonny Fuego's current shop.

◓

In most cases, the TIG welding process is favored for a strong frame. Here, Mike Flanigan of A.N.T. lays a bead after the frame has been tacked together in his jig.

Maestro Mechanics
Professional Custom Cruiser Builders

Jonny Fuego
Idaho's Chopper Creationist

In the wilds of lawless Idaho there is a bike builder who goes by one name: Fuego. His C&C Bikes (Cruisers & Choppers) is a basic, somewhat mobile operation: a heavily tooled garage that pops up in various corners of the Gem State. Wherever Fuego lands, strange two-wheeled creations soon appear.

Fuego is a prime example of the modern garage genius. Once he started tinkering, his destiny was sealed. At the age of eight he learned how to thread wheels with color-coordinating wire. When he needed to back up the claim to his name, he enrolled in welding school and mastered fire. Fuego approached an old western cowhand for an apprenticeship in the lost art of leather filigree; now his C&C Bikes feature custom embossed cowhide saddles.

Bikes remain a pastime for Fuego (his "real" job is in graphic design), but calling him a mere hobbyist would be like calling Picasso an enthusiastic doodler. Fuego has been building bikes in Idaho for nearly a decade now and has attracted a growing list of clients willing to spend thousands on hand-built and original designs.

The C&C Bikes workshop looks like a cross between a torture chamber and a mad machinist's laboratory. A massive medieval-looking device stands upright in one corner. This is a roll bender: one of several obscure tools used to create hand-built machines. On a rolling table nearby sits a rough suede bag filled with 175 pounds of lead shot. On this, Fuego will lay thin sheets of metal—soon to become faux gas tanks or crank covers—and forcibly coax them into smooth shapes using a blunt tool picked from a quiver of hammers.

Fuego is self-taught. He started with hand-drawn sketches of his vision. Then it was a trip to the hardware store for some plastic PVC pipe and duct tape. Fuego lit up his kitchen stove and fired those pipes to a soft and workable state. The nasty fumes, he said, were all part of the sacrifice. After seventy hours of hard labor, he had a life-size model. In 2002, he named his first frame the Hellbent.

The Hellbent is a classic chopper design: the pedals are well out in front of the low-slung saddle, giving the bike a recumbent feel while also freeing the rider to plant his feet on the ground. Around the same time Fuego made this innovation, Electra introduced their Townie series, marketed on the Flintstone car concept.

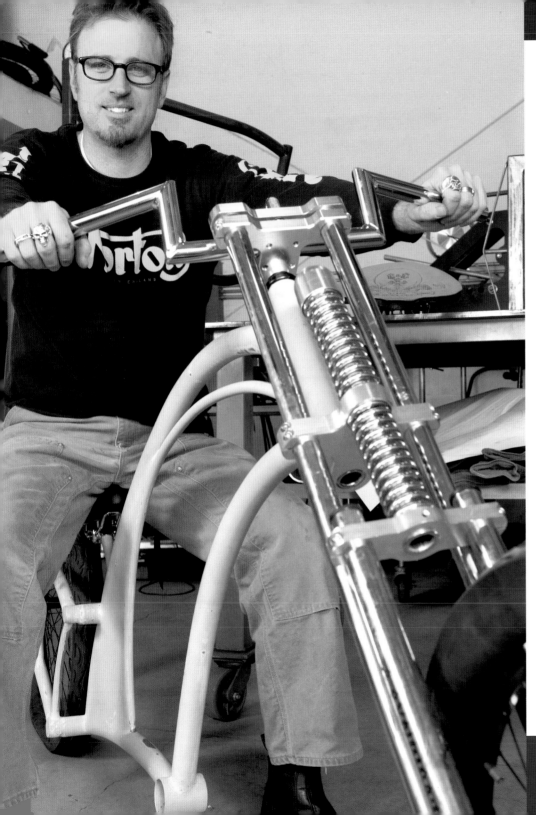

HammerGlide

Before the HammerGlide chopper gets set to paint, it is mocked up to make sure everything fits. Custom Z-bars and springer are attached to the front.

Following the Hellbent prototype, Fuego put himself through a rigorous self-education. He learned the ins and outs of bike building and eventually began utilizing graphic design software. Once digital, he transcended duct tape and melted PVC. He has since designed over a dozen distinct frames and translated these into actual rubber on the road. He receives quirky requests from neighbors and friends, including one from a man who asked for a cruiser with a sidecar for his toddler son. "People come to me and say 'I want something that says me. I want something that reflects who I am.'"

Ultimately, Jonny Fuego is a stylist. His life reflects his bikes and vice versa. He dreams of the day when he can just build. That's it—nothing more. Maybe in California. Or better yet, Mexico. Somewhere warm, where you cruise over to the taco stand for a cold *cerveza*. You tip it back and take a cold pull with your shades on, your feet on the ground, and your ass in the saddle.

Jonny Fuego

⊙ ⊙ ⊙

Grey Ghost

This is the first Fuego replica of a 1900's board-track racer. The tank is meticulously gas-welded, and then the seams are filled with bronze to look like a patch job done by a pit crew in the past.

↑ →

Grey Ghost

Before the engine and transmission are put in, the bike gets mocked up to check the fit. It can also be ridden as a regular bike.

Jonny Fuego

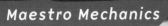

→ ⊗ ↓

HammerGlide

The seat is hand-tooled leather with a C&C logo, hand-stitched onto a fabricated seat pan. Check out the single-sided floating rear wheel, steel-filled frame, and twist throttle-style accelerator for the front disc brake.

Sam McKay
Firebikes: International Demand

"I'm pretty much a one-man show," Sam McKay says.

This might come as a surprise to the Firebikes fan base, a global assortment of admirers. His fans are also bike builders, people like Jonny Fuego, who cite McKay as a pioneer in the field of custom choppers. Today, Firebikes is a staple at trade shows such as the North American Handbuilt Bicycle Association and in megacruiser rides in several northern European cities.

The Firebikes story is an improbable one. McKay's Canadian hometown of Lumsden, Saskatchewan, is well over a thousand miles from the nearest surf break. Despite this isolation, McKay's humble hobby has grown into an internationally coveted brand.

There is a simple arithmetic to McKay's success: "I like building bikes because bikes have been a part of my life since I was five years old. When I was eighteen and all of my friends were driving cars, I was still riding bikes. When I was twenty-five, it was the same story." After retiring from BMX, he wanted to build something different and new.

In 1998, McKay struck creative gold when he built one of the earliest stretch choppers. Once his success launched, there was no turning back. After nearly a decade with Firebikes—a period during which he quit his day job to build full-time—McKay estimates he built five to six hundred bike frames. His dues thus paid, he turned to designing and drafting. Today, as a master of his craft, he lets apprentices do the dirty backbreaking framework.

This is not to say that McKay has become bigheaded about his skills. He is an equal-opportunity builder to the core. His only regret about success is how rising market demand for his bikes has raised his prices to astronomic levels.

"It's a little out of hand," McKay says.

High-end European clients are partly to blame. Firebikes has distributors in Sweden, Holland, and Germany, and nearly 80 percent of his total output is sold in northern Europe. As competitors and imitators sprung up in those countries, prices for authentic Firebikes surged.

McKay faced an issue of conscience. "I was getting a lot of e-mail from people who could not afford my bikes. You feel bad for the guy. He's been priced out. Now what do I do?"

Troubled by this unforeseen dilemma, McKay adapted by selling custom Firebikes components such as stems, sprockets, and pedals. Along with the parts, customers get a certificate of induction into the Firebike Army. Part sales now make up the majority of his business, and McKay's mind is at ease; he can focus instead on the perks of his success. When he travels to

Sam McKay

2008 Black Widow

McKay uses a bending machine to give his frames a signature gentle curve.

chopper-friendly Europe, his name precedes him. "If I go to Germany, I send a couple of e-mails before I leave and will have friends who come and meet me at the airport and show me around," he says.

McKay is grateful for his good fortune. "Who would have thought that these bikes would take me across the world?"

Morgitition

You can have Sam build it
with spokes or with custom
CNC'd machined wheels.

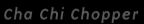

Cha Chi Chopper

Classic Firebikes 1.5-inch hand-rolled
steel frame, tig welded with custom
Firebikes accessories.

Eric Paulson
Freakbike Militia Mercenary

Picture south Florida's Gold Coast. What comes to mind?

Sand and sun: nice.

Hurricanes and gators: fine.

Geezers stooped behind the wheels of Oldsmobile Cutlass Supremes, drifting across lanes of wide, flat boulevards: terrifying.

Rib shacks. Sunburned tourists. A hundred and one things come to mind about south Florida before mountain biking.

Eric Paulson's story is tied to Florida's eccentric cycling community and starts with mountain biking in a place you'd least expect to find it.

Mountain biking is big in Florida. Or at least it used to be before the bulldozers cleared out most of the trails. Hyperdevelopment of open spaces is an issue across the nation. But in south Florida, where

Giant Killer

Handmade details include the chopper fork, frame, mock gas tank, seat, and, best of all, a skull headlight.

Lucky 13

Is this a motorcycle? No, but it looks natural sitting in the driveway.

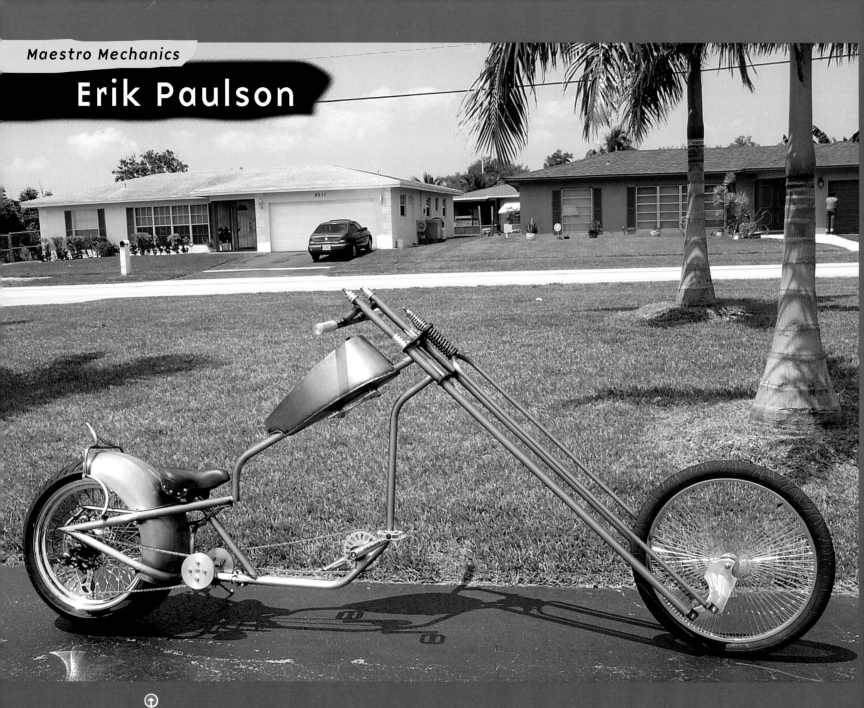

Erik Paulson

⬆

Blingatron

A long, stretched chopper with a motorcycle rear wheel, disc brakes, and multispeed gearing.

a narrow strip of solid land is bracketed by ocean on one side and swamps on the other, development takes a larger toll.

When Paulson moved to Fort Lauderdale from his native New England, one of the biggest surprises was finding an active mountain biking community. On weekends, devoted gear heads would gather on low wooded hills to build and ride tight and technical single-track trails. The hills were not natural features, but piles of dirt left over from decades-old construction sites. On these trails, Paulson consistently joined a core group of experienced mountain bikers, including world champion Marla Streb.

Those were the days before golf courses and strip malls devoured much of what was left of Palm Beach County's

unclaimed dirt piles. With their playgrounds disappearing, the mountain bikers needed fresh diversions.

Enter the chopper.

Paulson started by modifying production bicycles, and after a few years of polishing his skills, he evolved into bottom-up bike fabrication. "Certain people are just freaks for bicycles. I've always been like that," he says.

By day, Paulson is a ship builder, a job that helps reveal the secrets of large-scale mechanics. Bike building unlocks the same code. "When you are a kid, a bicycle is like a big mystery, it's something magical. You buy it and ride it, but you don't understand it," he says. Paulson thinks most people are intimidated by the often-simple engineering that makes things work. "Even adults think there's some factory with amazing machines that stamp these bikes out in one piece."

In the Palm Beach chapter of the Freakbike Militia (the largest order of this nationwide club—see chapter 4), Paulson is known as a real-deal craftsman. He has the tools, the shop, and the knowledge to forge high-end bikes.

"There's nothing like riding a bike you built. The sky's the limit—whatever you want to make, you can make it."

Pandora

An extreme custom cycle.

Erik Paulson

Pandora

Every inch of this masterpiece is hand-fabricated, machined, or tooled. The fork design is called a girder springer, popularized in the sixties.

→ ↓

Marcel Danek
Hardcore Cycles: When in Doubt, Build Your Own

Germans pride themselves on precise machinery. From Leica to the minds behind Electra Bikes, they are a people of mechanical prowess. No surprise then that Marcel Danek, a bike mechanic living outside Düsseldorf, is building radical and daring machines.

Early on, Danek was the type of Harley-Davidson groupie who found inspiration in American TV shows like *Orange County Choppers*. But when he decided to buy his own cruiser, he found only disappointments. Chopper-style custom bikes were not only overpriced, most were far "too normal" for his particular taste. So he did what any resourceful German would: he decided to build his own.

He searched for parts and learned that most were made in China of low-quality materials. Rather than give in to the corruption of free global trade, he decided to build his own components. He headed for the welding machines and the tube benders with a vision, "a bike like no other." After mastering the basic skills, he burst onto the custom cruiser scene with daring, unprecedented designs. He calls his trademark bikes "stretchcruisers": elongated frames with ridiculously long front forks and oversized wheels. He keeps his colors simple and lets the lines do the talking.

Eventually, Danek started his own bike-building business. Today, Hardcore Cycles sells customs and parts on the Internet and through Danek's store in Neuss,

Wild Thing

The long handmade frame and front end boast a wild animal print scheme.

⊙

How it begins: the mock-up stage is very important.

Marcel Danek

Germany. Through the Web, this once anonymous bike mechanic gained a widespread following; he is flooded with bike requests from Spain, Canada, France, Italy, and Dubai.

"I never thought that my parts and bikes would be public in the world," Danek says. Despite the success, he still tinkers and designs as a way to relax after work. He relishes the lifestyle of weekend rides with friends.

"Cycling is passion," he says. "The most important thing for a bike is that it is customized."

Hardcore Customs

A Woody design with 144-spoke wheels. Nice!

Marcel Danek

Hardcore Customs

A cute Hello Kitty one-off.

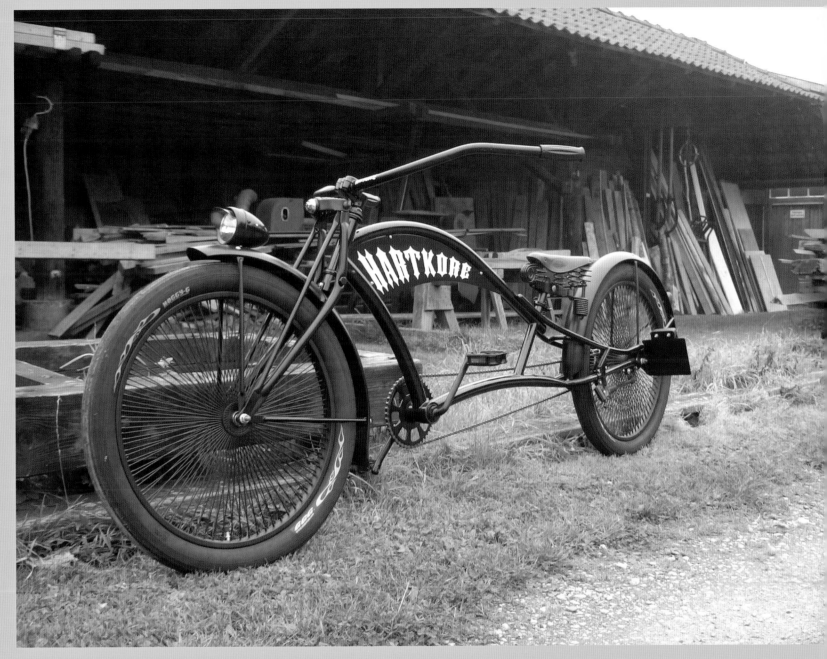

⬆ Hardcore Customs

Smooth, gentle curves and custom design keep Danek on top in the UK. The super-elongated frame gives this chopper a smooth ride.

Maestro Mechanics

Aaron Bethlenfalvy

Bethlenfalvy's designs have been published in countless periodicals and include numerous cover photos. He has received a "Best in Category" award from *Popular Mechanics*.

Concept Kiddie Trailer

Your little one can ride in style in this cruiser trailer. The body is hand-sculpted and encased in fiberglass. It features 100 percent leather tuck-n-roll upholstery, wall-to wall carpeting, a removable Carson top, a wolf whistle, and Hooga horns. And there's plenty of trunk space for diapers.

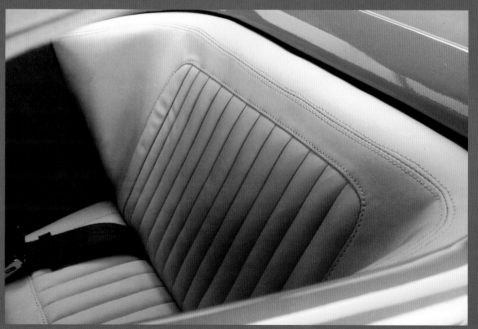

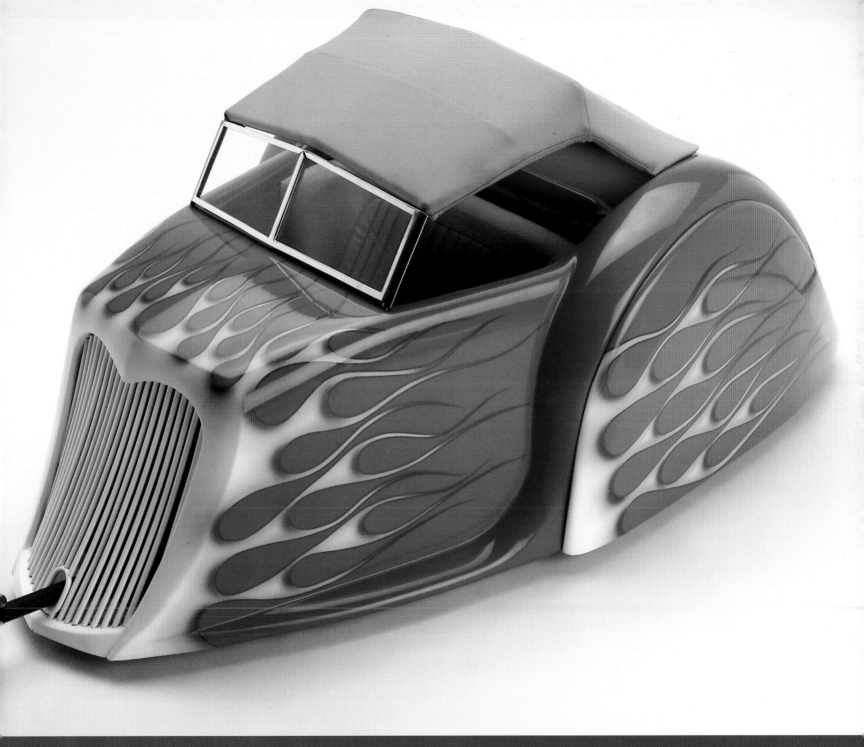

Aaron Bethlenfalvy

Chupacabra Chopper

As with any master custom, this is one of a kind.

A.N.T. Alternative Needs Transportation
Answering a Different Set of Desires

Mike Flanigan's journey reads like a bike odyssey. He has pedaled across the country and set up shops with little more than the clothes on his back. And in his time in the industry, with his earnest hard work, determination, and belief in quality bicycles, he has left his mark.

The foundation of Alternative Needs Transportation can be traced back to when he and his mother fled to the countryside in pursuit of an alternative, authentic lifestyle. They "started growing organic food, using wood heat, and kind of became redneck hippies," Flanigan says. Still just a kid, he became interested in renewable energies and ecological modes of living. In this emerging context, the bicycle became more than just a ride; it was a political statement.

Out in the world, Flanigan flirted with motorcycles, but his passion stayed with the simple human-powered bike. He led road tours, started bike clubs, and, after a nine-day cross-country ride, landed a job at Boston's Fat City Cycles. During his five years there, he fell in love with New England. In 1994, he helped launch Independent Fabrication, where he remained until 2002. A year later, A.N.T was born.

During this transition, Flanigan was instrumental behind a now-infamous guerilla marketing campaign directed at the owners of oversized SUVs. The inefficient, gas-guzzling trucks were targeted—usually overnight—with fake parking tickets that listed statistics about gasoline consumption. With a grassroots volunteer effort of one hundred walkers and cyclists, Flanigan helped ticket roughly ten thousand SUVs. The concept spread virally. Similar tickets were developed and distributed across the nation and even the globe.

Today, Flanigan's A.N.T. bikes are among the most artful hand-built cycles in America. They are practical—several baskets and cargo options are available—but also elegant. Through skilled hand-craftsmanship, high-quality materials (including wood and copper), and a passion that manifests in sublime design, Flanigan has situated himself among the elite bike builders working today.

Mike Flanigan turns fire
and steel into art with
beautiful welds and unique
components.

⊕ →

After completion, this bike
gets copper controls.

A.N.T. Bikes

Light Roadster

A great bike for commuting
to work.

Light Roadster Deluxe

The perfect combination of style and function.

A.N.T. Bikes

The bar-linkage brake system uses a friction plate that puts pressure on the top of the tire.

This is a very elegant bike, made with wood and copper lugs. Even the rims are wood.

Jason Coryer The Butcher of Buffalo

Southern California may be the epicenter of the cruiser lifestyle, but the skills and desires are not limited to the sun-drenched and bikini clad. In snowy, dreary Buffalo, New York, a hardnosed work ethic holds things in place, and classic bikes have always been, in the words of Jason Coryer, "a dime a dozen."

In Coryer's Buffalo, vintage two-wheelers are a piece of the landscape. "You can go to any grandpa or grandma's attic and find something cool,' he says. Coryer used to work at Rick's Cycle shop, the sun around which Buffalo's bikes orbit. Rick's is a century-old, tin-roofed, wooden-floored joint where you can walk right in and find a 1952 Schwinn Phantom sitting on the showroom floor. The guys who work at Rick's have a collecting problem—a four-to-five-bikes-per-week problem. At the height of his addiction, Coryer had thirty old bikes in his home.

He is a vintage aficionado, but rather than collecting and polishing pristine refurbished models, he creates vintage assemblies pieced together from aged parts. "I wanted something that these rich guys who buy a bunch of classics couldn't just buy," he says.

He finds bikes anywhere he can. He welded some wheel and chain guards onto a vintage JC Higgins he bought "from a guy who has barns and barns full of bikes." This bike became "The War Machine." Others are based on classic Indian or Scout motorcycles. When building a 1915 Cyclone replica, he collected photos from motorcycle museums to guide his process.

Coryer shrugs off the purists who take offense to his cavalier reassembly of vintage models. "The bikes are so common around here, I don't feel guilty about chopping up a prewar Schwinn," he says.

To finish a bike, Coryer does not shirk the details. His aesthetic leans toward an antique, weather-beaten look. To get several fresh coats of paint looking haggard is a challenge met with intentional rusting ("I throw it out in the rain and then paint a thin coat over the rust"), precise violence ("controlled chipping"), or simple sun-fading ("sticking a bike in a window facing the sun for about a year works pretty well").

Once finished, Coryer might sell his creations on eBay or to friends. "I don't like selling them. They're like my kids." Customers respect this protectiveness. Those Buffalonians lucky enough to ride his bikes through town still refer to them as Jason Coryer's bikes. Chopper bragging rights always defer to the creator.

Erik Tiles
Sun Valley's Cruiser Conquistador

Sun Valley, Idaho: playground for the rich.

In this town, where second homes sit empty most of the year, true locals are jaded about the things that money can buy. For this working community, showy material goods don't impress.

It's all about what you can build yourself. Modifying old cruisers is one way locals define themselves. The pastime comes in many forms. For some, it's enough to put a fresh coat of paint on a thrift-store special. But for the builder who knows the score, a custom bike is a way to bring design flair to the culture-starved northern Rockies.

Take Erik Paul Tiles. Architectural designer by trade, he built himself a reputation as a local bike artisan using little more than some tubes of grease and years of accumulated bike-shop know-how. In Sun Valley, a Tiles bike is easy to spot: it's the cruiser with bling.

Tiles builds customs for close friends. He won't take their money; a case of Coors tallboys will suffice. In a wealthy tourists' world of trophy homes and trophy wives, a Tiles bike is the authentic showpiece.

After building a handful of cruisers for friends, Tiles got down to the real business of his own ride. "I started with an idea, a big bright shining star," he says of the original vision. "I knew what it needed to be from the beginning." It would be "sparkly" and shimmer with reflective chrome.

First he needed a frame, and the one he settled on was as sad as they come. Unwanted and unloved, the orphaned bike was a Dumpster discovery. In Sun Valley, where the town thrift store is aptly named "The Gold Mine," it's not rare to find valuable toys in the trash. But Tiles's naked frame, encased in decades of rust, was true garbage. He took up some sandpaper and a wire brush. "That was the hardest work"; it took him ten hours just to strip the paint.

If some cruiser builders are mechanic freaks (Eric Paulson) and others are original design pioneers (McKay), Tiles is a fashionista. He begins with a bike that belonged to some far-off place that style never reached, to

Inside Tiles's cruiser lair.

Erik Tiles

someone with less fashion sense than a Wal-Mart greeter. He starts with this blank thing and creates something from nothing.

This is a feat of improvisation. While building his personal ride, he ran into a snag: some parts didn't fit. Specifically, the fork ring didn't fit onto his front end tube. Tiles peered around his dim garage shop for a tool and his eyes settled on an eighties-era vacuum cleaner. He picked up the extension tube, and, forcing the vacuum over the frame, took a big rubber hammer and began expertly smashing the parts into place.

"It's a lost art," he said with a smile, in between thwacks. It took Tiles roughly thirty hours to finish the Great Dane, a red-and-white-pinstriped homage to his alma mater, the University of Wisconsin, and Ron Dane, the Badgers' 1999 Heisman Trophy winner.

As the bike came together, Tiles was pleased. "This is the bike I wish I had had as a kid." By the time the Great Dane was a reality, it was dead winter in Sun Valley, his driveway covered in two feet of the white stuff. But after some hasty shoveling, he gave it a whirl. Riding his Great Dane up and down an icy driveway, smile frozen in place, Erik Tiles eagerly awaited Idaho's late spring thaw.

In the hills around Sun Valley, cruiser conquistador Erik Tiles rolls deep.

Josh Merrow Brooklyn's Project Bicycle

Josh Merrow's mechanical journey began with a stutter. His desire to build machines was pure, but there was a problem—a math problem. It was no particular equation or formula that had Merrow stumped, but all of mathematics.

Merrow was studying English Literature at the University of California at Berkeley when he finally faced his numerical demons. Berkeley's sculpture classes had access to vast World War II–era metalworking shops, and Merrow's brain ran in circles over potential machines he could build.

After enduring a couple of physics classes, he was back in the metal shop, running in a different sort of circle. His first mechanical design project was a human-sized hamster wheel. It stood nine feet tall. "I needed geometry to do that," he says.

After schooling himself, Merrow graduated to teaching. He moved to Brooklyn, New York, and found work as a special projects teacher at El Puente, a down-and-out junior high school languishing in the massive engineered shadow of the Williamsburg Bridge. The students were a teacher's greatest challenge: disinterested in academics ("a couple couldn't even read") but brimming with sometimes dangerous energy, these urban kids desperately needed a change that none of them could have named. Merrow, architect of human hamster wheels and his own tangential paths, was the right man at the right time.

"I cooked up a program for the kids who were flunking geometry," he says. "The idea was to engage them in another way." These kids were math casualties. "They had such humiliating experiences in math class—which I could totally relate to—that some of them would just lock up when they saw a fraction."

Merrow sought permission to take over a dingy, grassless backyard for a bike-engineering project. He started out with the most alluring aspect of design: imagination. "I asked them for their dream machines and they drew these exotic contraptions. Bikes with six or eight wheels and armchairs to sit in."

Of course, as reality crept in, the students' ambitions returned to earth. And as they learned the figurative and literal angles of building bikes, unique riding machines began to emerge from the gritty backyard at El Puente.

Merrow encouraged experimentation, often leading by example. As the students worked on their designs, he fabricated his own version of a classic Dutch women's bike, designed with long dresses in mind. It had a comically elevated saddle, recumbent pedals, and under-the-leg steering. Once complete, Merrow rode it fearlessly around the neighborhood. "I think these kids figured out that I was kind of a nut."

In the end, his eccentric approach created an enthusiastic movement and filled this corner of Brooklyn with an enduring local pastime.

Gecko Cruisers

John W. Lefelhocz at Cycle Path Bike Shop in Athens, Ohio, makes the Gecko Cruiser with 106,500 glass Czech beads, held in place by nylon thread, epoxy putty, and 493 feet of wire. By not using glue, he ensures the glass beads maintain a clear brilliance.

⬆ ➡

A seamless tessellated pattern, similar to the designs of M. C. Escher, is incorporated onto the tubes of the bicycle. Red and purple geckos come on the tubes of every bike, while blue, green, and yellow are used in combination to create a specific look.

Igor Ravbar Slovenia's Wood Craftsman

Every morning in Ljubljana, Slovenia, Igor Ravbar rides four miles to work on his own hand-carved wooden bikes. At the National Museum of Slovenia, he is an arms and armor conservationist, but at home, Ravbar is a woodworker. Before turning to bicycles, he carved kayaks and one canoe from wood. In May, 2002,

Woo 1

Woo 3 Ladies

he took Woo 1, his first all-wood bike, for a road test. Since that successful experiment, Ravbar has built two more entirely wooden bikes. The saddle, basket, chain guard, and fenders are all wood: incredible work that, one day, should be featured in the very museum where Ravbar works.

⊕ *Woo 2*

This recumbent bike is made from spruce strips and beech veneer. There is a two-part chain inside the wooden frame.

Olli Erkkilä Finland's Rising Star

Olli Erkkilä wanted to build things with his own hands. "First came bicycles," he says. When he had enough of that, he turned to motorcycles. In 2004, after years tinkering with bikes and motorbikes (and winning design awards in Scandinavian chopper shows), Erkkilä enrolled in Lahti University's Insititute of Design in southern Finland. Riding to classes on a "regular bicycle" was out of the question for this art student. After receiving positive feedback on his first bike—the super-stretched Funky Elephant—he set off in even funkier directions. Judging from the uninhibited creativity and skill behind recent designs like his Jopo Freak, it seems safe to say that Olli Erkkilä's design career is only getting started.

⬆ ➡ *Forkless Jopo Freak*

(←) (↙)　　　　　　　　　　　(↑)

2004 Funky Elephant　　　　　　*2008 Leaf Green Chopper*

Erkkilä's first custom design.

Olli Erkkilä

2007 '20s Bike

Another white-tired original
from southern Finland.

(←) (↑) (→)

Red Oxide Chopper

Super-narrow handlebars, sweet tires, and a recumbent feel make the Oxide one of a kind.

A Simple Guide to Customizing Your Ride

By Jonny Fuego

How hard is it to customize your ride?

A bicycle is one of the simplest machines out there: two wheels, a crank, and a chain. Dismantling it, buying the right aftermarket parts, and making a masterpiece—your own masterpiece—is simple. In this chapter I will customize two different bikes, each in its own way. As you will see, it's not as hard as you might think.

Concept: Bike in a box upgrade

Buying component parts off the wall at your local bike shop and slapping them on.

Tools

Adjustable wrench, flat-head and Phillips-head screwdrivers, and Allen wrenches. This bike was donated to this project by Ed Hickey at Phat Cycles and came shipped in pieces in the box. If you really want to learn about bikes, buy a bike in a box. Warning: this may void store warranties, as most bike shops like to put their bikes together for their own peace of mind and for your safety.

I built this bike in fifteen minutes. And as you can see, it's a pretty cool bike. But there are a few thousand of these on the road, and why should the one you ride look like everyone else's?

First, I go to the store and purchase $50 worth of parts that I feel will make for a cooler cruiser. It isn't a lot, but the outcome will be dramatic.

↻

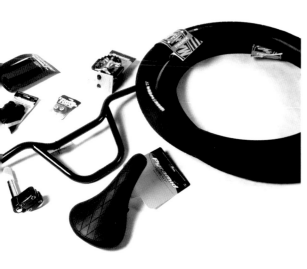

↻ Taking this cruiser apart is easy. And all the parts go on in a snap.

↑ ↻

Parts purchased and installed: BMX handlebars, grips, BMX stem (neck), fat slick tires, BMX foot pegs, BMX old-style diamond weave seat, BMX trap pedals, and 8-ball valve caps.

Custom #2

Concept: Hot bike from junkyard

Scavenging parts from a motorcycle graveyard, custom painting, and leatherwork.

Tools

Adjustable wrench, flat-head and Phillips-head screwdrivers, Allen wrenches, leather punch, basic leather tools, leather snaps, rubber gloves, disposable paint mask, plastic wrap, plastic bowl, auto adhesive pinstripes, small model paintbrush, pinstriping brush, black model paint, shades of tan paint, spray gun (or spray cans). WARNING! Follow all manufacturer's recommendations and warnings when painting, and use a well-ventilated area. Build at your own risk.

This is a theme build—slightly more intense, but still relatively simple. Like Custom #1, I will take apart a perfectly great cruiser and customize it from the ground up, with no structure modification. The theme is leather and steel.

This bike was brought to me by Chad and was perfect as-is except for the two flat tires and dust. In my town, there were five of these exact bikes running around throughout the hood, and sure, it's cool and all, but it was just like all the others.

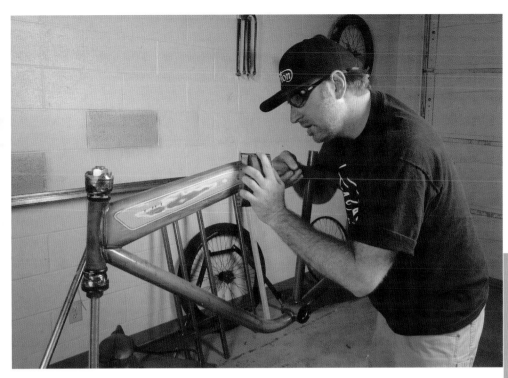

Using automotive-grade masking tape, I mask off the area to paint. I also go to a sign shop and have them make me a mask for a logo that I will add to the headtube.

After removing all the parts, I bolt the bike to the table using a stand I built from a threaded steel rod bent at a 45-degree angle. I then secure the stand to the bike frame using two bolts and washers that fit into the headtube.

I give the bike frame a good sanding with 400-grit wet/dry paper.

Then I wipe it down with a degreaser and clean off the dust.

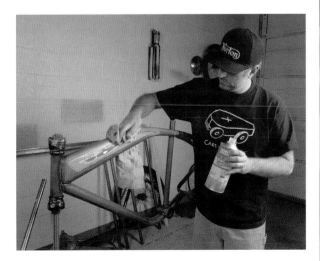

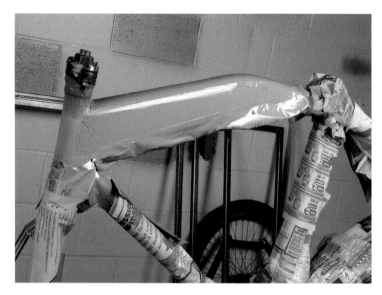

I tape newspaper on the surrounding bars and spray primer on the areas I want to paint.

With a spray gun, I apply a base layer of tan to match the leather I will be adding later. The colors don't have to match exactly—just get them close. You can use spray cans instead of a spray gun, but be sure to wear a paint mask and rubber gloves.

From a hobby store, I purchase a few small bottles of tan model paint darker than the base layer I have sprayed. I pour the paint into a plastic bowl, and with a balled-up piece of plastic wrap, dip and apply as shown. Do not glob it on; use light pats of paint to ensure a smooth finish.

Like this.

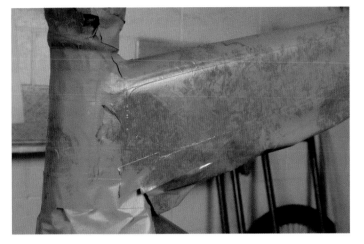

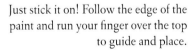

After the paint is completely dry, I remove the masking and newspaper.

For this project I will show you two ways to pinstripe. This is the easiest way.

Just stick it on! Follow the edge of the paint and run your finger over the top to guide and place.

I use a small model paintbrush and a bottle of black model paint to outline the logo on the front (headtube) and I have the same sign shop cut me a graphic for the side.

With a pinstriping brush and a bottle of pinstripe paint, I add an accent of red. You can also cheat and use an adhesive stripe, which is much easier.

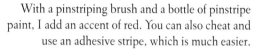

Using a can of clear coat, I lay five coats down over everything. Remember to wear a paint mask and rubber gloves.

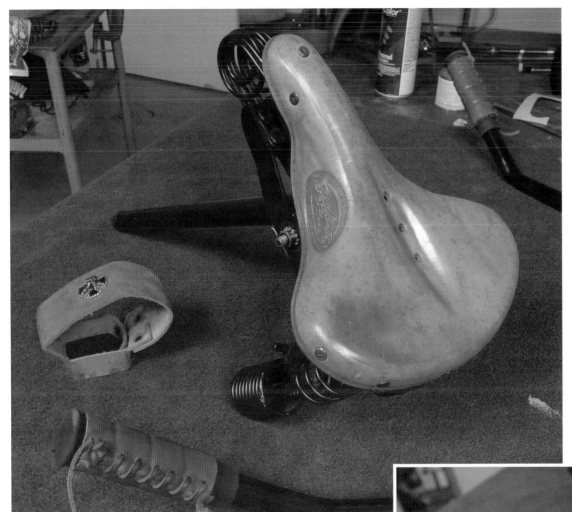

 Leather! Ed at Phat Cycles sent me this very cool back-stock leather saddle, to which I match the color of leather and make a stem (neck) cover. The grip covers are just a piece of flat leather cut to fit around the grips, with the holes punched and laced as shown. For the stem cover, I use a piece of paper as a template and fit it, then transfer it to leather, cut it out, and add snaps.

Here are the grip covers.

Grip covers, stem cover, and a cool set of "Z" bars scavenged from a motorcycle graveyard.

Cross pedals and an RPM Sprocket.

I made a hub polisher as well.

No one in the neighborhood has this!

Bicycle Culture, Clubs, and Cults

Something is afoot in America today. People are back on bicycles. They are pedaling to work to save money on gas. They are riding to the grocery store, home from parties, or out on a sightseeing tour of an American city. Bike-love groups are gathering for rides, drawing the curious attention of pedestrians and the enthusiastic support of town councils. Coast to coast, bikes and bike riding are coming back with such momentum, cities and towns are challenged to meet the demand. Bike paths and bike racks can't be built fast enough. Portland, Oregon, has set a bike-transport model for American cities to emulate. The speed with which this organic movement is sprouting has taken many by surprise. America's bike reawakening is in high gear.

Alleycat

Go to any major city in the United States or Europe and you'll likely encounter bicycle messengers. Since the mid-1980s, cities around the world have relied on the contemporary incarnation of bike messengers to deliver packages efficiently. Their selling points include their ability to maneuver through traffic faster than cars, and to navigate the back roads, alleys, and shortcuts around the city, which allow them to deliver packages unencumbered by the hassles that accompany driving.

There are significant messenger populations in Boston, New York, Chicago, Montreal, Berlin, London, Toronto, and San Francisco. These cities are home to frequent races that test messengers' skills against each other, from the "Stupor Bowl" in Minneapolis to "Beer Not Bombs" in San Francisco. There are World Bike Messenger Championships, held in a different location every year. And there are smaller races put on in messenger communities as well. Known as "alleycat races," these events traditionally pit messengers against each other to ride through the city and visit various designated stops. These underground races are not sponsored by corporations or sanctioned by organizing bodies, but are put on by messengers for other riders.

Alleycat races are intended to simulate the everyday activities of bike messengers. Riders gather at a designated starting point. None of them know the route of the race beforehand because there is no set route. Instead, riders are given a manifest that lists all of the stops they must visit to collect verification from a designated point person. Often the stops will have specific tasks that riders must complete. Tasks can range from drinking a particular beverage to picking up an unwieldy package to pumping up a flat tire. There are no rules, and the rider is free to map their own course. The first rider to reach a finish point with a completed manifest is the winner.

—Katy Dang

Most messenger bikes are "Fixies," where the rear hub fixed to the cog means no coasting. If the wheels move in a single direction, the crank and pedals move with them. Most Fixies do not have brakes. If you need to slow down, your legs do all the work.

Charleston's Cruiser Ambassadors

Charleston, South Carolina, is a slower-paced kind of destination, a place where a New Yorker might go to unwind and smell the salty air. At the Vendue Inn, they feel that cruiser bikes are the vehicle for visitors' relaxation. For over ten years, the Vendue has offered cruiser bikes to guests during their stays.

"It's nice for them to be able to truck around town," says Cat Morgan, human resources manager at the Vendue. Morgan says weekends are ideal bike time in Charleston. Many visitors arrive in cars and expect to see the town from behind car windows. When they pull up to the Vendue and see a rack of sea-foam-green cruisers, a lightbulb goes off.

Wait a minute! the tourist will say. *I don't need to start my car, to burn more gas, to sit trapped in passive travel! Instead, I can see this town at my own pace, turn my vehicle with the power and balance of my own body.* This is an exciting moment, a lightning bolt of clarity for the suburban prisoner of the road. It's a lifestyle choice, a reconnection with the outdoors, with self-reliance, and with one's surroundings. By offering these bikes, the Vendue Inn invites its guests to make a small but substantive change in their lives.

South Carolina: always ready in soul.

The Freakbike Family

In south Florida, a man named Kenny Prather used to enjoy mountain biking with his friends. When the trails they built were replaced by one housing development after another, Prather and company were robbed of a pastime (see "Customs" chapter). They loved riding together, but it seemed society was literally devouring their freedom to roam.

Faced with this challenge, Prather inverted his mindset. Why run away from development, on trails winding through jungle overgrowth, when the streets beckoned? Why give up bike riding when it had always brought him so much joy? From these simple questions, Prather's necessity bore the fruit of invention.

You would figure that, as the creator of something called the Freakbike Militia, Prather would be a revolutionary, or at least a subversive type. But he is a family man who makes sure to put his kids to bed before tinkering with his bikes. By day he does good deeds working as a hospital maintenance mechanic. As a law-abiding, American-values type of citizen, Prather admits that he's not the stereotype for the movement with the "kinda scary" name.

It all started on a Palm Beach Halloween. Prather and his old mountain biking posse hung some flyers and started spreading the word. They were putting together "a ride." No one knew exactly what would happen. A love for bikes was the only criterion. That first ride drew forty-seven people, on all kinds of funky cruisers and customs, many of them fished from local trash bins.

"It's amazing what people put in the trash," Prather says. He found barely used BMX bikes and other bikes in top condition except for a flat tire or two. Mechanic ignorance leads to waste, and Prather was set to transform consumer trash into fun and creative diversions.

The Freakbike Militia favors wild experimentation. "Stretch the frame, the forks, bend the tubes, lower the seat, flip the frame, make

⬆ Nerd night brings people, bicycles, and a theme to the streets.

⬅ Kenny "Clean" Prather builds outrageous simple machines. Here, he poses alongside his seven-foot bike with, of course, its own pirate flag.

it a tall bike, paint modifications, make it stand out," Prather chants. Anything to make a bike more freaky. "Make it as wild as you can, as long as you can ride it."

A Freakbike Militia ride is not a marauding gang. These are not the Hell's Angels. As much as these bikers like to adopt the image of revolution, their gatherings are peaceful. Businesses stay open late on ride nights; onlookers line the sidewalks to watch the mild mayhem. Host cities are enthusiastic. In Palm Beach, the militia's epicenter, the mayor even joins in on the fun. In Delray Beach, golf-cart retirement land, the city makes sure that the tribe comes to town for the Christmas parade. Roughnecks and pillagers these are not.

Standing as it does for no single race, creed, or class, the Freakbike Militia has become a sort of rolling democracy. People of all ages and backgrounds join Prather, Eric Paulson, and the rest of south Florida's cycle creationists.

"We have kids who are seven and eight years old and then a seventy-two-year-old man riding right next to them. We have lawyers, mechanics and even a judge," Prather says. In 2007, a pirate-themed ride drew over 190 swashbuckling cyclists. The militia has become an equalizing force in a society otherwise segregated by status.

Since 2003, Freakbike has spread virally. At last count, there were ten chapters in North America. Prather visits them to see how his message has been translated. In Ontario, Canada, Prather says, the bikes tend to be better kept and more cared for. For every chapter, the militia rides are a refuge from work, routine, and the rigidity of society. Freakbike rides offer a different sort of community and a freedom that breeds creativity and friendship.

At Freakbike Militia group rides, one must ask: Is it the bikes or the people that are freaky? All walks of life and all types of bikes come out for this ride.

Nerds and bikes: a perfect combination?

Some serious customs come out of hiding for Freakbike rides.

Arrgh! Pirate night!

Many events have major sponsorship that believes in the bicycle lifestyle. Participants in this "bike rodeo" show off their fun and wild customs.

The Future

At the Electra bike company, plans are being made for the future. Benno Baenziger believes that America's transportation culture is due for a change and he wants to capitalize.

This was a year when Americans took a good long look at themselves. We took note of our staggering rates of material consumption and our oil-guzzling ways. Looking into the mirror, we saw imperfections, and in certain small ways we worked to fix ourselves up, to make ourselves just a little bit better.

This is part of what feels good about getting back on a bicycle. Sure, we can recycle or buy low-wattage lightbulbs. But riding a bike has a compound dividend: we feel good about saving on gas and not contributing to traffic congestion, but we also have a healthy, honest good time in the process.

America is a big country. We are not, nor will we ever be, Holland or Belgium. But we are ultimately a nation of towns and cities, and for the increasing numbers who believe in low-consumption lifestyles, bike riding has taken on new significance.

It is with these beliefs in mind that Electra introduced the Amsterdam line. The idea this time is not so much about the bike itself, but about the lifestyle changes these simple machines afford and encourage. The Amsterdam bikes are attractive, but ornamentation is an afterthought. In an ironic circular path, the company that brought style and form back to the bicycle industry has returned to simple functionality. Baenziger: "I think the one thing I would like to see is people living and working and existing in neighborhoods where it's fun and safe and easy to ride a bike around."

G Men's Classic
3 Amsterdam

We stand at a crossroads—vaguely aware that our habits as consumers and commuters need to change, though unsure how. It is possible, and indeed likely, that bicycle use will rise to unprecedented levels. A trend that swelled with fashion and style may now surge into a fast-moving wave powered by a more time-honored American trait: pragmatism.

The bicycles of our near future will be practical. They will wed usefulness to hipness and function to expression. America can embrace a reborn bicycling culture. The next American century will return home, to local values. Our towns and city parks, our storefronts and front yards will again play host to the free motion of bicycles built to last.

Acknowledgments

There are too many people to count who have supported the creation of this book. If we have forgotten you, please forgive us. *Cruisers* could never have been made without you.

Particularly, we would like to thank Jennifer Pattison Tuohy for giving life to this idea way back when it was an article in the *Sun Valley Guide*. Special thanks to our talented photographers, Chris Pilaro, Sonya Prather, and Piper Loyd. Thanks also to Benno Baenziger for answering questions in the middle of the night from Taiwan; Tim Brandt at Bicycle Chronicles; Aaron Bethlenfalvy; Sam McKay; Freakbike's Kenny Prather and Eric Paulson; Brett King at Felt; Gary Silva; James Traub; Dave Stromberger at Nostalgic.net; Erik Tiles; Josh Merrow; Marcel Danek; Olli Erkkilä; Jason Coryer; Leon Dixon; the friendly folks at the Vendue Inn; and cruiser models Casey Mills and Doug Oplt.

Resources

Collectors and Historians

Chris Lockhart
   Lookn4bikes@msn.com

Classic Rendezvous
   www.classicrendezvous.com

Marc Pfisterer
   easywind2@yahoo.com

Dave Stromberger
   dave@nostalgic.net
   www.nostalgic.net

Kristofer Nyström
   knystrom@purdierogers.com

Tim Brandt
   www.bicyclechronicles.com

Custom Builders and Designers

Aaron Bethlenfalvy
   Director of Industrial Design
   www.pacific-cycle.com

Mike Flanigan/A.N.T Bike
   (Alternative Needs Transportation)
   www.antbikemike.com

Sam McKay—Firebikes
   www.firebikes.com

Marcel Danek—Hardcore Customs
   www.danek-hardcore-cycles.de

Jonny Fuego
   www.jonnyfuego.com

Danu Huber
   www.hukkbikes.com

Kenny Prather
   www.freakbikemilitia.org

Woodpeckers
   hazeldon.woodpeckers@
   btinternet.com

Olli Erkkilä
   olli.erkkila@gmail.com

Ravbar Igor
   igor.ravbar@nms.si

Bicycle Companies and shops

Felt Bicycles
   www.feltracing.com

Phat Cycles
   www.phatcycles.com

Electra Bicycles
   www.electrabike.com

Dyno
   www.kustomkruiser.com

3G Bikes
   www.3gbikes.com

Bob's Bikes
   www.bobs-bicycles.com

Capitol Schwinn
   208-336-2453

Raleigh
   www.raleighusa.com

Photographers

Chris Pilaro—Photographer
   Cap13@mindspring.com

J Hopkins Photography
   Jen Hopkins
   jenhop@gmail.com

Sonya Prather
   www.sonyaprather.com

Piper Loyd
   www.piperloyd.com

Photo Credits

3G: 68

Aaron Bethlenfalvy: 73, 98–103

Bob's Bikes: 43–44, 47 bottom, 50–53, 67

Chris Lockhart: 12, 16–17 center, 17 right, 20 left, 24, 25 top, 28, 30–31, 33–35, 39

Chris Pilaro: 4–6, 8–9, 41, 71, 74, 77, 111

Danu Huber: 6, 25 bottom

Dave Stromberger: 15

Dyno: 48–49

Eric Paulson: 86–91

Felt Bicycles: 2, 54–66, 70, 160–161

Gecko Cruisers: 115

Igor Ravbar: 116–117

John W. Lefelhocz: 115

Jonny Fuego: 78–81, 125, 126–135

Kenny Prather/Freakbike: 142–155

Ranell Nyström: 36

Marc Pfisterer: 1, 13–14, 16 left, 18–19, 20 right, 21–23, 26–27, 29, 32, 37–38

Marcel Danek: 92–97

Mike Flanigan: 75, 104–109

Olli Erkkilä: 118–123

Piper Loyd: 3, 11, 112, 113

Raleigh: 47 top

Rico Shen: 69

Sam McKay: 83–85

Sonya Prather: 142–147, 149–151, 154, 156–157, 160

Vendue Inn: 140